DATE DUE			
		.	

16 - 19 00

523.7 Bredeson, Carmen. YGC3056618
Bre
Exploring the sun

**A.J. SMITH ELEMENTARY LIBRARY
UNION SPRINGS, NY 13160**

438569 01900   17441C   0001

A.J. SMITH ELEMENTARY LIBRARY

Exploring the sun

LAUNCH INTO SPACE!

Exploring the SUN

Carmen Bredeson
with
Marianne Dyson

Enslow Publishing
101 W. 23rd Street
Suite 240
New York, NY 10011
USA

enslow.com

Words to Know

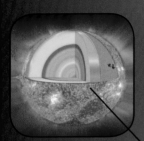

asteroids—Large rocks that orbit the Sun.

comets—Big balls of dust and ice in space.

core—The center part of something.

cosmic—Having to do with the universe.

energy—Light and heat.

orbit—Moving in an oval path around a star, planet, or moon.

solar—Having to do with the Sun.

temperature—How hot or cold something is.

Contents

Words to Know 2
The Center of the Solar System 4
A Star Is Born 8
So Far Away 10
The Color of the Sun 14
Here Comes the Sun 18
Sun Safety 22
Sunshine 26
Day and Night 28
Learn More 31
Index 32

The Center of the Solar System

The solar system is the Sun and everything that orbits around it. Earth and other planets, moons, asteroids, and comets all orbit the Sun. The Sun is in the center of the solar system.

The Sun is a star. A star is a big ball of burning gas. From Earth, other stars look like tiny lights. Our Sun looks big because it is much closer to Earth than other stars.

Cosmic Fact

It takes eight minutes for sunlight to reach Earth.

A Star Is Born

Stars are made when gas and dust in space form a giant spinning ball. As the gas is pressed tighter and tighter, the ball gets very hot. The gas turns into energy. The energy comes to Earth as heat and light.

Cosmic Fact

The Sun was born nearly 5 billion years ago, long before Earth was formed.

So Far Away

The Sun is 93 million miles (150 million kilometers) from Earth. To reach the Sun from Earth, you would have to drive a car that is going 60 miles (97 kilometers) an hour for 177 years!

The Sun is a medium-sized star, but it is much bigger than Earth. More than a million Earths could fit inside the Sun. Medium-sized stars burn their gases more slowly, so they last longer than big stars.

Cosmic Fact

If the Sun were the size of a basketball, Earth would be the size of a pinhead about 1/16 of an inch (2.2 millimeters).

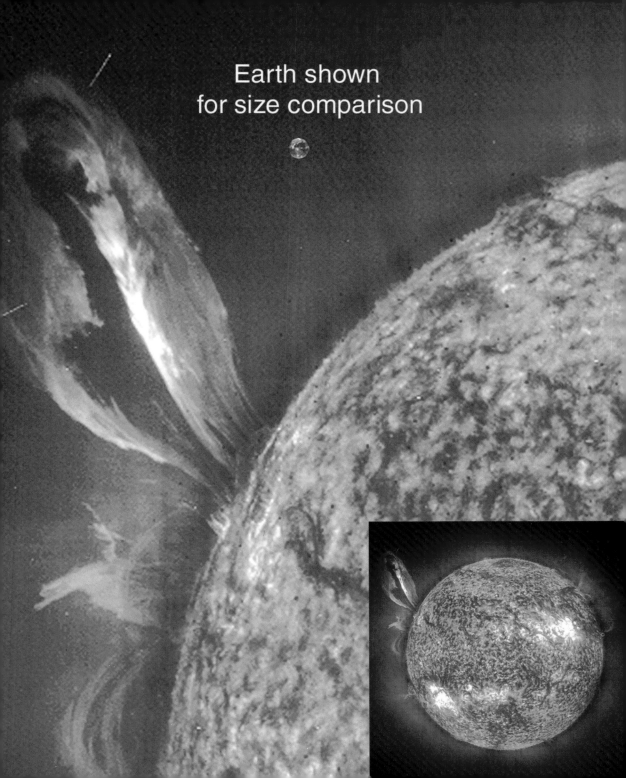

The Color of the Sun

Stars are many different colors. Very hot stars look blue. The coolest stars look red. The Sun is just the right temperature to look yellow.

The Sun is made of layers of gas. It does not have hard ground like Earth. You could not stand on the Sun's surface. It would be like trying to stand on a cloud. It is also very, very hot!

Cosmic Fact

The hottest part of the Sun is its center, called the core.

Here Comes the Sun

The Sun keeps Earth warm. Without the Sun, plants would not grow. People and animals would freeze. The Sun gives us the light, heat, and food that we need to live.

Cosmic Fact

Plants turn sunlight into food.

Some buildings have solar panels on their roofs. The panels trap sunshine and turn it into electricity. Solar electricity is clean. It does not make the air dirty. Ships in space also use sunlight to make power.

Cosmic Fact

Solar panels turn sunlight into power.

Sun Safety

People, plants, and animals need sunlight to live. But too much sunlight can hurt your eyes and burn your skin. Sunglasses, sunscreen, and clothing help protect people from too much Sun.

The Sun is too bright for your eyes. Never look directly into the Sun, even when you are wearing sunglasses. The Sun can make you blind! Scientists use large telescopes to look at the sun.

Sunshine

Scientists believe that the Sun has been around for nearly five billion years. It is about halfway through its life. That means it will probably shine for another five billion years.

Cosmic Fact
Sundials were the first clocks. They use a shadow from the Sun's light to tell the time of day. Sundials do not work at night.

Day and Night

Where does the Sun go at night? Nowhere! The Sun stays in place. It is Earth that moves. Earth turns all the way around each day. It is day on the side of Earth facing the Sun. It is night on the side of Earth facing away from the Sun.

Many scientists study the Sun. Astronomers use special telescopes to learn about the Sun. Aerospace engineers make spaceships that fly around the Sun. Solar engineers design buildings that use energy from the Sun. And computer scientists keep track of the temperatures on the Sun. Maybe you will study the Sun someday!

Learn More

Books

Jackson, Tom. *Magic School Bus Presents: Our Solar System*. New York: Scholastic, 2014.

Jemison, Mae. *Exploring Our Sun*. New York: Scholastic, 2013.

Simon, Seymour. *Our Solar System* (revised edition). New York: HarperCollins, 2014.

Web Sites

kidsastronomy.com
Research resources, games, activities, and fun facts on the exciting world of outer space.

nasa.gov/audience/forkids/kidsclub/flash/
Learn about space, space travel, astronauts, and read stories by other kids.

Index

asteroids, 4
comets, 4
day, 26, 28
distance to Sun, 10
Earth, 4, 6, 8, 10, 12, 16, 18, 28
electricity, 20
energy, 8
light and heat, 8, 18

moons, 4
night, 26, 28
planets, 4
plants, 18, 22
solar power, 20
solar system, 4
stars, 6, 8, 12, 13, 14
temperature, 14, 16, 30

Sun
age, 8, 26
color, 14
protection from, 22
size, 12, 13
sunburn, 22
surface, 16
telescopes, 24, 30

Published in 2016 by Enslow Publishing, LLC.
101 W. 23rd Street, Suite 240, New York, NY 10011

Copyright © 2016 by Carmen Bredeson

Enslow Publishing materials copyright © 2016 by Enslow Publishing, LLC.

All rights reserved.

No part of this book may be reproduced by any means without the written permission of the publisher.

Library of Congress Cataloging-in-Publication Data
Bredeson, Carmen, author.
Exploring the sun / Carmen Bredeson with Marianne Dyson.
    pages cm. — (Launch into space!)
    Summary: "Discusses facts about the Sun, including its place in the solar system, its birth, what it is made of, its size, why it is yellow, how animals and plants use sunlight, and what kinds of scientists study the Sun"—Provided by publisher.
    Audience: Ages 8-10
    Audience: Grades 4 to 6
    Includes bibliographical references and index.
    ISBN 978-0-7660-6835-3 (library binding)
    ISBN 978-0-7660-6833-9 (pbk.)
    ISBN 978-0-7660-6834-6 (6-pack)
    1. Sun—Juvenile literature. I. Dyson, Marianne, author. II. Title.
    QB521.5.B743 2015
    523.7—dc23
                        2015000144

Printed in the United States of America

To Our Readers: We have done our best to make sure all Web site addresses in this book were active and appropriate when we went to press. However, the author and the publisher have no control over and assume no liability for the material available on those Web sites or on any Web sites they may link to. Any comments or suggestions can be sent by e-mail to customerservice@enslow.com.

Portions of this book originally appeared in the book *What Do You Know About the Sun?*

Photo Credits: Christos Georghiou/Shutterstock.com, pp. 2 (solar system), 5; Dimedrol68/Shutterstock.com, p. 23; Elenamiv/Shutterstock.com, p. 19; EpicStockMedia/Shutterstock.com (Cosmic Fact background); ESA & NASA, p. 13; Jess Alford/Stockbyte/Getty Images, p. 25; Milanares/iStock/Thinkstock (blue/green background throughout book); NASA, pp. 2 (Sun diagram), 17; NASA, ESA, and the Hubble SM4 ERO Team, p. 15; NASA/JPL-Caltech/UCLA, p. 9; Otmar Smit/Shutterstock.com, pp. 2 (solar panels), 21; Revenant/Shutterstock.com, p. 11; silverjohn/Shutterstock.com, p. 27; Valeri Potapova/Shutterstock.com, p. 26; xfox01/Shutterstock.com, pp. 1 (Sun), 3; zhudifeng/iStock/Thinkstock, p. 7; Zmiter/Shutterstock.com, p. 1 (rocket illustration).

Cover Credits: Milanares/iStock/Thinkstock (blue/green background); xfox01/Shutterstock.com (Sun); Zmiter/Shutterstock.com (rocket illustration).